What Do Scientists Do?

Daniel Jacobs

Scientists look...

and listen.

Scientists measure...

and take notes.

Scientists draw plans to help them make new things.

Scientists take photographs to show what they found.

They also make charts
to show what they found.

We are scientists, too.
We look at things.

We measure...

and take notes.

We watch how things change.

We draw pictures.

We ask questions and work together to find out answers.

We make charts to show
what we find.

We are scientists every day.